Whales

Published by Wildlife Education, Ltd.
12233 Thatcher Court, Poway, California 92064
contact us at: **1-800-477-5034**
e-mail us at: **animals@zoobooks.com**
visit us at: **www.zoobooks.com**

Concept developed by KIDesign. Scientific Consultant: Mark Rosenthal, Curator Emeritus, Lincoln Park Zoo, Chicago, IL.
Photos: Cover, Stephen Frink Collection (Alamy Images); Half-title Whale, Royalty Free (Alamy Images); Title Whale, Tom Brakefield (Getty Images); Pages 2-3, Ken Usami/Photodisc Green (Getty Images); Page 4 Top Left, Noel Hendrickson/Photodisc Red (Getty Images); Middle, Tom Brakefield/Photodisc Blue (Getty Images); Bottom, Tom Brakefield/Photodisc Blue (Getty Images); Page 5 Top, Vince Streano (Corbis); Bottom, Stephen Frink Collection (Alamy Images); Page 8 David Fleetham/Taxi (Getty Images); Page 9 Top, Peter Steiner (Alamy Images); Middle, Royalty Free (Dorling Kindersley); Bottom, David Young-Wolff/Stone (Getty Images); Pages 10-11, Wolfgang Polzer (Alamy Images); Page 12 Top, SCPhotos (Alamy Images); Bottom, David Fleetham (Alamy Images); Page 13 Top, DLILLC (Corbis); Bottom, Brandon Cole/Visuals Unlimited (Getty Images); Pages 14-15, Digital Vision (Getty Images); Page 14 Top, Bruce Herman/Stone (Getty Images); Bottom, Sally Holden (CERF Photo); Page 15 Top, Photos.com.

ISBN: 978-1-932396-45-4

Whales

Wild, weighty, watchful, waterproof...

Ww

"Whales" starts with the letter "w". How many "w's" can you find on this page? Can you think of any other words that start with the letter "w"?

warm-blooded, well-mannered, world-travelers...

Wonderful, wonderful, wonderful, wonderful, whales!

I think "wet" would fit here, too!

Whole Wide World

There are two types of whales and they eat different things. **Toothed whales** have teeth, which they use to catch fish and other sea creatures.

Beluga whales live in the cold waters near the North Pole. These smiling whales are born dark gray or brown, and turn white as they get older.

Did you know that **dolphins** and **porpoises** are whales, too? They're the smallest members of the whale family!

Killer whales, also called orcas, are known to hunt for food in packs just like wolves. These whales have sharp teeth to catch and eat fish, seals, and even other whales.

4

written by Agnieszka Biskup

# of Whales

Baleen whales don't have teeth. They have bristly, comb-like baleen plates that work like giant filters. The whale gulps a mouthful of ocean, then uses its tongue to push the water out through its baleen. What's left are all the tiny animals and plants that were in the water. Yum!

The huge **gray whale** is an incredibly strong swimmer. Each year it swims from its breeding grounds off Mexico to its feeding grounds in Alaska, a round trip of about 12,000 miles.

Humpback whales have very long front flippers. The males are famous for singing long, complicated underwater songs. You can even get their songs on CD!

The One and Only

The world's oceans are full of living things. Swimming fish, swimming reptiles, swimming mammals, even swimming birds! Underwater, life is everwhere. But the whale is the ocean's biggest, most fascinating animal. It is truly one of a kind.

illustration by Carla Kiwior

There are 5 orange fish in this picture. Can you find them? There is only **1** yellow fish. Do you see just **1** of anything else?

Wondering About

Get your wet and wild whale questions answered here!

Is a whale a fish?
Whales live in the ocean, but they're definitely not fish. Whales are mammals, just like squirrels, lions, bats, and many other animals—including people!

Can you think of **1** other mammal that lives in the ocean?
Can you think of **1** other animal that's bigger than a car?
Can you think of **1** other animal that eats fish?

written by Agnieszka Biskup

Whales?

How do whales breathe?

Whales breathe air, just like people, so they have to come to the surface to breathe. Instead of nostrils, whales have blowholes on the tops of their heads. When a whale dives underwater, it closes its blowhole, just like you might plug your nose. When it first surfaces, it breathes out, making a spout.

So . . . just how big *are* they?

Some whales are pretty small (one toothed whale is only three feet long), but others are *huge*. A blue whale can be 100 feet long—that's about as long as three school buses or a basketball court!

BUS STOP

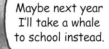
Maybe next year I'll take a whale to school instead.

How to Sing with Whales

by John Bowen

It isn't hard to sing with whales
But many people fail
Because they try to sing along
Without first speaking "Whale."

Instead, they wail their screechy songs
While standing by the sea.
They warble at the wind and waves,
"Wah-ricki, wicki, weeeeeee!"

When whales hear this, they swim away
And laugh, "Why, that's the worst!"
Because to sing with whales, you see,
You have to listen first.

How do they sing without opening their mouths?

This poem has lots of "w's" in it, though you might not see them all at first. Take a closer look. Can you find the word **whales** 4 times? Can you find the word **with** 3 times?

Now can you find these "w words," too?

without, when, why, worst, whale, wail, while, warble, wind, waves, wah-ricki, wicki, weee

Here are some tricky ones, with the "w's" in the middle! Can you find them?
Bowen, swim, away

Whales do a lot more than just swim in the water. Let's go whale watching!

Watching Whales

Breaching

Splash! Whales leap out of the water, then make a big splash when they come down. No one's sure why whales breach. It could be a way of saying "Here I am!" or maybe it's just plain fun.

Spyhopping

Sometimes, a whale will stick its head straight up out of the water to take a look around, as if to say, "What's up?"

written by Agnieszka Biskup

Yipes! That certainly got my attention!

Lobtailing

Slap! Sometimes, a whale will slap its tail flukes hard on the surface of the water, making a very loud noise. Perhaps it's trying to get someone's attention?

Logging

With all the leaping and tail-slapping they do, sometimes whales need a break. They rest by lying still, floating on the surface of the water.

The Long Journey

by Rachel Young

Every spring, gray whales swim thousands of miles north from the warm waters of Baja, Mexico to the icy waters of Alaska. In the fall, they swim back south again. But one small group of whales seemed to stay put, spending the whole year off the coast of Canada.

Scientist William Megill began to wonder if they really did stay in one place. To find out, he took pictures of whales in Canada in the summer. In the winter, he took pictures of the gray whales he saw in Mexico.

When he compared the pictures, he found that some whales had been in both places after all! These Canada gray whales were taking the same long trip as their Alaska cousins.

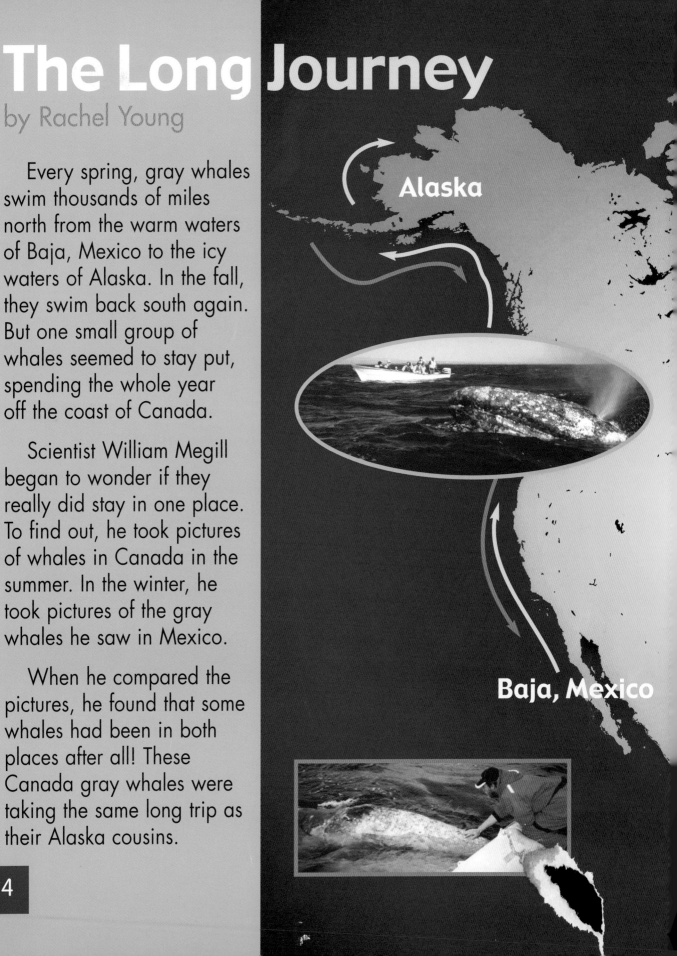

Alaska

Baja, Mexico

You Can Be a Scientist, Too!

Many animals, including songbirds and monarch butterflies, travel to warmer places in the fall and winter. These yearly journeys are called migrations.

Does your family take the same vacation every year? Maybe to the beach or to Grandma's house? Find a map that shows where you live and the place you visit. Trace your path from one place to the other. What are your favorite things to see along the way?

Ever wonder what animals pass through your area on their migration journeys? Have an adult help you use the library or the internet to find out, then look out for migrating animals in your backyard or park.

Flip-flopping and Spyhopping

by Donna Latham

In the deep blue sea, Clara and Skip enjoyed a whale of a playtime. Whoosh! Head first, Clara burst out of the water. She tossed her body into the air, flip-flopped, and arched her white belly toward the sky. Smack! Her black back whacked the waves. She bobbed in the tickly bubbles that danced around her.

Now, it was Skip's turn to show off. He zipped through the water, circled Clara, and nudged her playfully. Then Skip shot a bumpy flipper out of the sea and slapped the surface of the water. Crrrack!

illustrations by George Angelini

Suddenly, Skip bumped Clara and zipped away. He swished his tail fluke up and down as he swam faster and faster. Clara flapped her long, skinny flippers to turn around. She scooted after her friend. The chase was on!

As they played, Clara and Skip wandered away from their pod. Above them in the water, something rocked with the waves. What could it be? Always curious, Clara swam toward it. Swoosh! She sprayed a misty, v-shaped spout from her twin blowholes. Then she poked her knobby snout out of the water and popped up for a look. She spied something strange bob, bob, bobbing on the sea.

With her reddish-brown eye, Clara peeked curiously at the boat. A little girl squealed with excitement and pointed.

"Look!" cried the girl. "That whale is watching *us*!"

The captain said, "It's spyhopping. That's when a whale peeks out of the water like a periscope on a submarine."

Clara dove beneath the waves with a splash. She zipped through the water, circled the boat, and brushed up against it gently. The little girl giggled and clapped as Clara swam alongside the boat.

Clara felt Skip's familiar nudge. He was reminding her that she had drifted too far from the pod. Clara flip-flopped in the water and waved her tail fluke at the boat, then smacked it against the sea with a playful splash. Then she dove back under the surface and swam back toward the pod with Skip, as the little girl waved back after her.

The Adventures
of Otto and Allie

story and art by Meg McLean

Detective Wilkins & the W-Bandit

Detectives ask all kinds of questions—but the most important are **Who**, **What**, **Where**, **When**, and **Why**.

The **"W-Bandit"** has struck again, and Detective Wilkins is on the case. If the detective knows the **W-Bandit** only likes things that begin with the letter **"w"**, can you help him figure out **Whodunit**?

WHEN did it happen?
- ❑ Monday
- ❑ Tuesday
- ❑ Wednesday

WHERE did it happen?
- ❑ A museum in Miller, Massachusetts
- ❑ A ewe farm in Wiggum, Washington
- ❑ A cave in Cartersville, California

WHAT did the bandit steal?
- ❑ Woolly Paw-Warmers
- ❑ Snappy Sneakers
- ❑ Books about Bears

WHO is the bandit?
- ❑ Anthony the Alligator
- ❑ Hildegard the Hippo
- ❑ Wanda the Wallaby

art by Mo Ulicny

Now put a **"w"** in each blank to find out **WHY** the bandit did it!

She ____anted something ____armer to ____ear!

Ww

Whales features the letter "w," and developmentally speaking, it's a great letter. The "w" sound is one of the first sounds children learn, and it is rarely mispronounced. ("R" may come out as "w," but not vice versa.) Plus, the upper- and lowercase versions look alike, preventing confusion.

Zootles®
Resource
Corner

1

As parents, how often do we say, "just **1**" or "**1** more time"? We probably say the number **1** more than any other number, and because it's the first number, **1** is usually an easy concept to master. The only difficulty is getting that index finger to stick up straight. (In fact, most kids start counting with their thumbs. Try it. It's much easier!)

Who's WHO?

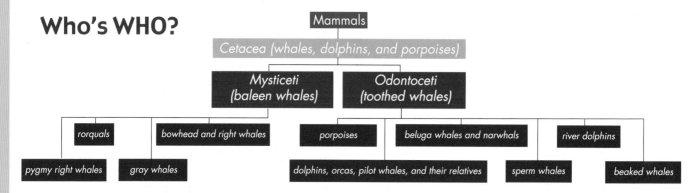

Mammals

Cetacea (whales, dolphins, and porpoises)

Mysticeti (baleen whales) — **Odontoceti** (toothed whales)

rorquals — bowhead and right whales — porpoises — beluga whales and narwhals — river dolphins

pygmy right whales — gray whales — dolphins, orcas, pilot whales, and their relatives — sperm whales — beaked whales

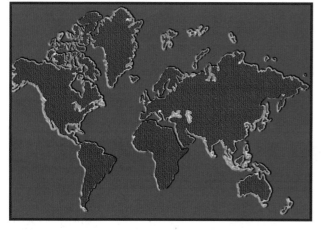

Where in the WORLD?

Whales live in all of the world's oceans. A few cetacean species also live in freshwater rivers and lakes in North America, South America, and Asia.

● **Whales**

Reading Resource

Every title in the *Zootles* series is designed to be used for fun and learning, and as a reading resource as well. The pages are written simply and address various stages of emerging literacy, and they encourage new readers to exercise their new skills at just the right level. Reading *Zootles* together will provide "together time" for you and your child—and reinforce vocabulary, comprehension, and early reading skills, too.

It's a ZOO out there!

Getting face-to-face with a whale is an awe-inspiring experience. Some whales, particularly dolphins, can be observed at aquaria, zoos, and theme parks, but a trip on a whale-watching boat is even more thrilling, because you will likely see the animals in their natural habitat. Day trips are available along the entire Pacific Coast as well as in New England and eastern Canada. Wherever you watch whales, help your child be a good observational scientist:

● How can you tell what kind of whale it is?
● Do you see behaviors such as breaching, spyhopping, and lobtailing?
● How is the whale responding to the humans who are watching?

What ELSE can we DO?

- **Wonderful Whales (p. 2-3)** Poets talk about consonance, the repetition of one sound to create a poetic effect. Elementary school teachers repeat one sound to teach children phonemic awareness. Repeating sounds brings brain-based learning and poetry together!
  - **ZOOTLES TO-DO:** Don't limit yourself to the **w** sound— we have more than thirty other sounds in English. Play around with others. Choose another sound—maybe the first sound (not necessarily the letter) in your child's name—and think of words that begin with that sound.

- **The Whole Wide World of Whales (p. 4-5)** Baleen feels a bit like plastic. Both whales and women suffered when baleen was used to make the corsets that gave 19th century ladies their hoop skirts and hour-glass figures!
  - **ZOOTLES TO-DO:** Help your child understand how baleen works by making some "whale food." Put some water in a bowl and add whatever you find in the cupboard that won't dissolve: leftover cookie sprinkles, coarse salt, rice, uncooked noodles, dried beans, etc. This is the "ocean" a whale sucks into its mouth. Now pour your ocean through a colander. See how the whale pushes water *out* of its mouth while baleen keeps the good stuff *in*!

- **The One and Only (p. 6-7)** Whales are part of a highly complex ocean ecosystem. Help your child identify the different creatures in *The One and Only*.
  - **ZOOTLES TO-DO:** Help your child draw his own ocean scene using crayons (not markers) on heavy paper. What things could he add that aren't in *The One and Only*? Paint over the entire picture with blue water colors or tempera paint. The paint won't stick to the crayon, creating a super underwater effect.

- **Wondering About Whales (p. 8-9)** The concept of mammals can be a bit confusing for preschoolers. Remind your child that mammals are animals that—like us—grow hair, have live births, and nurse their young.
  - **ZOOTLES TO-DO:** While some whales can only hold their breath for a few minutes, a sperm whale can hold its breath for about an hour. Using a stopwatch or the second hand on the clock, help your child time how long she can hold her breath. How long can she hold her breath after she's jumped up and down twenty times?
  - **ZOOTLES TO-DO:** How big is big? Head out to the sidewalk with chalk and a tape measure to compare a blue whale (100 feet) to a humpback whale (50 feet) to a dolphin (8 feet). You can even compare a whale to an elephant (20 feet).

- **How to Sing With Whales (p. 10-11)** After thirty years of study, scientists are able to answer many "who," "what," "when," and "where" questions about whale songs, but they still don't know much about "why." And so far, no human can really "speak whale."
  - **ZOOTLES TO-DO:** Whale songs can be heard at many locations on the internet. Sit down with your child for a listen and then join in the chorus. You can ask for an encore performance at bathtime.

- **The Long Journey (p. 14)** Human children learn through imaginary play. Playing "house" helps them understand family relationships. Playing "store" helps them understand the broader world. By playing "whale," they might learn some complex scientific concepts, too.
  - **ZOOTLES TO-DO:** Try playing "migration." You can spend summer (a few moments) in Alaska (the bedroom), but when the days get short, start swimming down to Mexico (the living room) with plenty of adventures on the way. When the days get long, head back to Alaska, singing your whale songs as you go.

- **Flip-Flopping and Spyhopping (p. 16-19)** Clara and Skip show us the similarities between our own children's behavior and that of young humpback whales. They all like to play with friends, explore their environment, and wander back when they head too far afield.
  - **ZOOTLES TO-DO:** Your child can make his own Clara or Skip by stuffing a brown paper sack with crumpled newspapers. Tie off a tail with a rubber band, and paint the bag to look like a whale. Then head off to explore the mysterious ecosystems of the backyard.

> Otto, do you think this counts as hair?

Allie: An intrepid hedgehog

Otto: An adventure-loving otter

Zootles®
Resource Corner